RAILWAY LIVERIES

LONDON MIDLAND & SCOTTISH RAILWAY

Brian Haresnape

LONDON

IAN ALLAN LTD

Important Note

The Author will be pleased to receive letters and photographs, or informative descriptions, related to this series. Items accepted will be retained and paid for at standard rates on publication; those he is unable to use can only be returned if they are accompanied by a stamped addressed envelope, otherwise they will be filed for possible future use.

Material, either commissioned or freely submitted, is provided at the contributor's own risk and neither the Author of this book, nor the Publisher, Ian Allan Ltd can be held responsible for loss or damage. Colour pictures should be carefully protected against damage if sent by post.

These books will eventually cover a large part of British railway history, from the pre-Grouping liveries to the present-day BR Corporate image. Private individuals, or societies, who have made a study of any one railway company's liveries and who would appreciate the opportunity to have their researches published at no cost to themselves, are invited to contact the Author, c/o Ian Allan Ltd, Terminal House, Shepperton TW17 8AS, England. All contributions will be duly acknowledged in the narrative. Every care is taken to credit material used and no responsibility can be held for error or omission.

1 *Title page inset*
The coat of arms, or crest of the London Midland & Scottish Railway Company, in the form applied to rolling stock and some road vehicles and publicity. *British Rail*

2 *Title page*
Crossing the Border. The up 'Coronation Scot' express at speed near Gretna Green, hauled by Class 7P Pacific No 6223 *Princess Alice*, in the 1937 livery of blue and silver. *W. Hubert Foster*

First published 1983

ISBN 0 7110 1281 4

Published by Ian Allan Ltd, Shepperton, Surrey; and printed by Ian Allan Printing Ltd at their works at Coombelands in Runnymede, England

Contents

Introduction

The LMSR Image

It was called the 'LMS', but the full title of 'London Midland & Scottish Railway' was LMSR in abbreviated form. And as such I will refer to it in this text; otherwise I will find myself writing about the 'GW', the 'LNE' — both plausible — and the S — quite ridiculous!

The LMSR was a huge railway; the biggest of the 'Big Four' created in 1923 and it took time to weld the various constituents of the new company into a cohesive whole. One of the most important contributions towards creating a corporate image was the adoption of a recognisable LMSR livery, which would replace the various colour schemes of the absorbed pre-Grouping companies. Perhaps the LMSR directors were aware that the other three of the 'Big Four' were selecting green, of differing shades, for their locomotives, or perhaps it really was an all-powerful Midland Railway bias, that led them to select the MR crimson lake as their colour scheme — not only for locomotives but for carriages as well.

Many myths and legends attend the early days of the LMSR, and their choice of a livery is no exception. In particular it has been alleged by many writers that the former LNWR Crewe works did not like the idea of painting its locomotives in MR colours. This was not really the case, and there is ample proof that Crewe was actually quick to paint some of its more important express passenger locomotives, as well as some smaller types in the new colours, (see photo for example). The reason that so many ex-LNWR types retained their black livery during the 1920s was the method employed by Crewe for repair work, which had become highly organised and efficient, partly under the guidance of H. P. Beames. Locomotives receiving intermediate repair, or specific alterations, or collision damage repair were not necessarily repainted; but 'patch-painted' where necessary. Black it should be emphasised is the easiest of all to 'patch-paint'. It was both effective and economical. By way of contrast, to change the complete livery of a locomotive from black to crimson lake took time, and was more expensive, and was thus restricted by Crewe to major overhauls, or new construction, which included a period of days in the paintshop. Locomotives receiving light repairs at other workshops on the system, such as Rugby or Bow, were not repainted, although sometimes the existing insignia was retouched.

Indeed the cost of applying the crimson lake complete with lining-out and final varnishing evidently worried the LMSR directors, because after some six years it was decided to use the colour only for the more important passenger types and to replace it with varnished black with a single red line for 'intermediate' or as we nowadays call them, 'mixed traffic' types. Goods engines were in plain black throughout from 1923 to 1947.

The colour, crimson lake was the same shade as that used on the former MR, and indeed to begin with the same style of gilt and black numerals and gilt shade of lining was employed. During the latter part of the 1930s, when the lining had been changed to chrome yellow and the lettering was often in chrome yellow shaded red, it was suggested that the crimson lake *appeared* to be darker, but to some extent this may simply have been an optical effect created by the brighter numerals and lining.

There is another possible explanation for this suggestion that the crimson was darker. By good fortune some colour film was taken of LMSR locomotives in the late 1930s, and even allowing for fading and weak colour rendering, which makes some of them unsuitable for reproduction, these authentic photographs show the crimson lake to have a definite tendency to darken and blacken after some time in service; especially on those areas exposed to heat. No doubt locomotives were not so frequently and meticulously polished as had been the case in MR days, so the blackening effect steadily built-up, in particular on the boilers and fireboxes. There is no reliable evidence to show that the LMSR ever changed the shade of the crimson lake in prewar days, but this blackening effect may well have led to the suggestion that a darker shade was being used.

Until the outbreak of World War 2 the LMSR had a very strong and attractive image, with the added colour and appeal of the recently introduced streamlined express trains. The latest Stanier Pacifics graced the most important duties and the Stanier rolling stock was modern and comfortable. The initials 'LMS' were recognised by the public immediately, although the company also put its full title on many poster boards and station signs, and it used the circular coat-of-arms or crest on a wide variety of applications, from carriage sides to posters.

3
The prewar LMSR scene in its heyday, when all the locomotives and rolling stock were liveried to a basically uniform style. The date is 22 September 1938, and the location is Camden Bank, outside Euston. Fowler 'Patriot' Class 5XP 4-6-0 No 5525 *Colwyn Bay* climbs the incline with the 4.35pm express to Birmingham and Wolverhampton whilst in the background a Euston-bound suburban electric train waits at signals. All the rolling stock in the picture and the locomotive itself are finished in the crimson lake livery. The destination boards on the main line carriages are white, with black lettering. *E. R. Wethersett*

3

The war years had a particularly severe effect upon the LMSR, due partially to the important bomb targets of the industrial areas it served, and the system was in decidedly poor shape when peace was restored, with a huge backlog of repairs and maintenance for locomotives and rolling stock and track, and a surprising amount of bomb-damage to repair on its buildings and structures. The directors abandoned all hope of a quick return to prewar speeds and schedules and as a result, the 'Coronation Scot' streamlined train was never reinstated. During the war the crimson lake had been abandoned for locomotives, to be replaced by overall plain black, whilst carriages only retained the crimson lake if 'patch-painted' (which most were) otherwise they were given a plain *maroon* finish. This maroon *appeared* to be of a bluer shade — ie more purple — than the true Midland crimson lake, and it was more opaque in quality because probably a single coat had to suffice; this could have been achieved by the addition of a little white. Locomotives did not receive it, except a few 'patch-painted' examples where areas such as the cabside were dealt with; some even lasting into early BR days and carrying BR numbers; the last one recorded being Compound 4-4-0 No 40934 still in crimson lake until September 1951. Only *one* locomotive was officially repainted maroon after the war, as an experiment, and this will be described later.

The name maroon is therefore I suggest appropriate to the paint used in the later war years and afterwards and which was almost certainly of different quality and manufacture to the fine shade, known as 'Midland Red' (the crimson lake) of prewar days. What is not 100% certain is whether

4

4
The postwar LMSR scene, by contrast, was typified by black locomotives and even the streamlined 'Pacifics' were given this drab finish, as seen here on Class 7P 4-6-2 No 6243 *City of Lancaster*, photographed on a Perth-Euston express approaching Leighton Buzzard tunnel. This locomotive was destined to be the last to remain in service with a streamlined casing, being rebuilt in June 1949, by which time it was running with the BR No 46243 but still in LMSR black livery. *W. S. Garth*

5
A very early suggestion for a new company livery for the LMSR, tried-out on a Hughes ex L&Y 4-6-0 No 1670 and photographed in full 'photographic grey' treatment. The lining style was pure L&Y but the letters LMS in simple sans serif form upon the tender were new. The locomotive retains its 'Horwich 1923' combined numberplate and worksplate on the cabside. The basic livery would presumably have been black. *British Rail*

the actual shade was deliberately altered to look 'bluer', or whether it was *intended* to be the same colour. What caused the change of shade was probably simply the requirements dictated by wartime shortages and economies, namely that carriages and locomotives no longer received such careful preparation and undercoating as had been the prewar practice. The opaque maroon allowed quick top coating, whereas the prewar paint was more translucent and relied to an extent upon the colour of the final undercoat to give it its intensity. In 1946 the LMSR started to use the name maroon officially, for carriages and for the lining on locomotives, and BR perpetuated this name.

My own personal recollections and observations of the LMSR commence with the middle war years of my childhood and result from almost daily 'train-watching' from a convenient lineside recreation ground near Kenton on the ex-LNWR main line. The later streamlined Pacifics made a lasting impression, with glimpses of their crimson and gold colours sometimes showing through the grime that seemed to cover everything. Particularly impressive to my young eyes was the bright shade of vermilion red applied to the bufferbeams of the recently repainted black locomotives — it seemed almost orange, and was certainly more visible than the 'signal red' that BR adopted in its place. I was also fortunate enough to see the last 'Claughton' 4-6-0 No 6004, still in shabby vestiges of prewar crimson lake, and working freight trains from Willesden. Also recalled are a number 'Royal Scots' and 'Jubilees' in similar condition; normally only recognisable as such by their hastily cleaned numerals and lettering (a common practice was to clean just the cabsides) which revealed the crimson lake beneath the grime. But the vast amount of locomotives that powered their way past were in black, and filthy dirty towards the end of the war. About this time my 'spotting' expeditions took on a double role, watching the skies for flying-bombs ('doodlebugs'), as well as the trains!

The final years of the LMSR are portrayed in part two, including the 1946 livery experiments which resulted in the decision to retain black for locomotives. This was a sad choice which reflected the mood of these austere times, and it was with some pleasure that I observed the various shades of blue and green that BR experimented with in 1948/49 — but that will be the subject of a later book in this series!

This series of books is written with the model maker very much in mind, and eventually the entire historical span of British railway liveries will be covered. Within this format it is not possible to go into exhaustive and minute detail about specific livery applications; the aim is to present a broad but concisive description of the livery applications and policy changes of each railway, thus enabling the reader to research still further if so desired, armed I hope with sufficient background information.

Because space does not permit there are some omissions which should perhaps at least be mentioned here. In particular, I have not described the rolling stock of the narrow gauge Leek & Manifold Light Railway, which the LMSR took over from the North Staffordshire. This, and the LMSR's two tramway systems — the Wolverton & Stony Stratford Tramway and the Burton & Ashby Light Railway — did not survive very long. In at least two cases, the Leek & Manifold and the Wolverton & Stony Stratford, the stock (or some) received crimson lake livery. Also worthy of mention is the Manchester South Junction & Altrincham electric rolling stock, which was of LMSR design but which had a *green* livery (fully lined) because it was a line jointly owned with the LNER. Also for space reasons I have not described the special liveries applied by the LMSR to post office sorting carriages, with their large 'Royal Mail' lettering, and the remarkable insulated 'sausage vans' operated for Palethorpes. The sight of one of these, with the huge slogan 'Palethorpes Royal Cambridge' along the length of the crimson lake bodysides, and with

5

6
The first year or so of the Grouping witnessed some curious hybrid liveries such as the one carried on this former North London Railway 4-4-0T, which became LMS No 6466. In LNWR black, with full lining-out, Bow Works has simply removed the cast numberplate from the centre of the sidetank and applied the Midland Railway style numerals, in unshaded pale yellow, or gilt. No LMSR coat of arms, or company initials were carried. *Ian Allan Library*

a highly realistic picture in full colour of a pack of sausages, (about 4ft×4ft 6in) was one not easily forgotten! These special vehicles are all described in *The LMS Coach 1923-1957*; a book referred to later.

The practice of putting the locomotive power classification on the locomotive cabside should be mentioned. This was derived from MR days when brass numerals were used. In the LMSR period the style of painted or transfer numerals and letters, eg 5XP or 4F, followed the style of the main insignia as a rule. Two other details, the cast iron smokebox numberplates and the oval cast iron shedplates were also very characteristic touches; their style and location being readily deduced from photographs, as they were usually picked-out in white paint, and I have not included detailed close-ups of these features.

In the 'Preface to the Series,' which appears in *Railway Liveries Southern Railway* (published 1982) I have given the reader a basic introduction to the subject and to the problems which exist in any attempt to describe livery schemes of the past. In this third book of the series I feel it necessary to repeat some of these observations, and I beg the indulgence of those readers who have already encountered them.

Although the emphasis of both text and illustrations is understandably upon the liveries of the locomotives and rolling stock — the figureheads of the railways — there are other aspects which help to create a wider spectrum and I have decided to include architecture (mainly stations), road vehicles, publicity and some other elements to a degree at least offering a general idea of what occurred at any given period. The specialist in search of deeper knowledge of these associated elements is referred to the excellent work being done by the members of many societies at the present time. In particular I would mention the Historical Model Railway Society and the Railway Correspondence & Travel Society, both of whom publish very detailed information upon railways, their stock and their liveries, from time to time. In this instance, the LMSR has also been remarkably well researched and documented in recent years by members of the LMS Society and by two noted experts in the field. Information in great detail is contained in their two books which are thoroughly recommended to the reader: D. Jenkinson's *Locomotive Liveries of the LMS* (Roundhouse Books/Ian Allan) and R. J. Essery and D. Jenkinson's *The LMS Coach 1923-1957* (Ian Allan) already mentioned. On the specialist subject of goods vehicles a recent book by R. J. Essery and K. R. Morgan, *The LMS Wagon* (David & Charles) gives much useful detail.

Finally, a word on the subject of colour illustrations. With so many preserved railways in existence today, and with an array of operational rolling-stock restored to what purports to be its original livery, or style, a strong temptation existed to take a camera and record these in colour for use throughout the series. For two basic reasons I have decided not to do this. One is that wherever contemporary evidence of a livery exists, I prefer to use it, as it is of course completely authentic even allowing for the known vagaries of film emulsions and suchlike. The other reason is that, sad to say in a fair number of cases, what is otherwise truly

7

7
Crewe Works locomotive maintenance policy whilst under the guidance of H. P. M. Beames was extremely efficient and well organised and no time was lost in repairing locomotives and returning them to revenue-earning service. One aspect of this was that many locomotives and tenders would go through the overhaul without complete repaint — particularly the black engines — and a patch-painting policy was adopted. This led to such mixed images as this one, where the tender of 'Rebuilt Claughton' 4-6-0 No 5953 *Buckingham* still has LNWR lining-out, despite the date of the photograph, 1928, by which time the second LMS livery style, had been applied with the initials LMS replacing the numerals on the tender, and these numerals painted on the cabsides. *Ian Allan Library*

8
Another example of Crewe's thrifty painting policy is seen on 'Rebuilt Experiment' 4-6-0 No 5554 *Prospero*, photographed at Coventry about 1929. In plain black livery, (there are some faint traces of LNWR lining on the tender, in the original photograph) the locomotive carries an LMS standard smokebox numberplate and the numerals are painted on the cab sidesheets. The tender however clearly shows a blank patch in the centre, where the numerals were previously located, and which have simply been obliterated with black paint. *A. Flowers*

8

excellent restoration work is spoiled by inaccurate livery details — perhaps due to lack of professional painters, or lack of proper lining-out and final varnishing. In the worst instances some of these preserved items are in the wrong colour, or shade, for the period which identifies the mechanical condition of the locomotive or carriage concerned. There is no point in perpetuating errors or omissions of this kind. Happily there are a good many really authentic museum-restored pre-Grouping locomotives and rolling stock in existence, and for this period, prior to the availability of colour film, I will be making some use of them. For the Grouping (certainly from mid-1930 onwards) a considerable amount of original colour material has been unearthed in recent years, in particular by the diligent efforts of Ron White of 'Colour-Rail', and by the now defunct 'Steam & Sail', and wherever possible I will use this rare and interesting collection. Some of these early colour photographs have faded, some have the blues and magentas too evident, but even allowing for this they bear the stamp of authenticity and carry with them a fine air of nostalgia.

Today various methods exist whereby colours are standardised and classified by numbers. There is, for example the BSS (British Standard Specification) range, and the well known Munsell method. Within these ranges many of the colours used for railway liveries can be accurately matched, but I have chosen not to refer the reader to these directly. Instead, each book in the series will have a colour sample range, matched by the printer as accurately as possible (allowing for the difference of luminosity and depth between printing inks and paint) and these will be numbered, with a prefix denoting the railway concerned. (In this instance the crimson lake, or 'Midland Red' as it was also called is ref: LM1.) All references in the text will refer the reader to the colour chart which appears in Ernest F. Carter's book *Britain's Railway Liveries*, published by Harold Starke Ltd. This chart contains no less than 50 colours all pertaining to railway liveries, and it is quite unsurpassed as a reference work. The text of the book, alas, is not so comprehensive and does not clearly define many of the variations, as it consists of a collection of references concerned with liveries culled from contemporary magazines and other sources. This limits its value, because many of the contradictory statements reprinted therein are not resolved, but rather left to the reader. The range of colours produced by Mr Carter's own research work are, however, quite superb. These are referred to in this series by the prefix C followed by the number on the colour chart; eg LM1/C28 is my reference for crimson lake.

I am grateful to A. B. MacLeod and the Ian Allan Library for much assistance and advice during the compilation of this book, and to John Edgington of the National Railway Museum, and S. W. Stevens-Stratten, for additional assistance. Ron White of Colour-Rail has been, as always, most helpful in locating contemporary colour photographs — no easy task! I would like to emphasise that readers are most welcome to add informative comment or criticism, addressed to: the Author, c/o Ian Allan Ltd, Terminal House, Shepperton, Middx TW17 8AS England. Specific personal correspondence cannot be entered into, but all relevant contributions will be acknowledged with thanks. (Please see the 'Important note' at the beginning of the book.) Finally, every care has been taken to credit the illustrations used and no responsibility can be held for error or omission.

Brian Haresnape FRSA NDD
Ramatuelle, France.
September 1982

9

9
A clear example of patch-painting, or more accurately, 'retouching' is shown by the numerals and lettering on Class 2F 0-6-0T 'Dock Tank' No 7166. The existing letters and numbers have been retouched to make them more visible; the locomotive itself has not been repainted. This was a very common sight on the LMSR from the war years onwards, and certainly existed long before then; being basically a Crewe practice. This detail is from a photograph of the mid-1940s.
Ian Allan Library

1: The Prewar Years 1923-1939

It was in October 1923 that the directors of the newly formed London Midland & Scottish Railway Company announced that they had decided to paint all its locomotives (except goods engines) in the crimson lake livery of the former Midland Railway. The goods engines were to be plain black. The first locomotives to appear in the 'Midland Red', or crimson lake (LM1/C28) were reported to be an ex-LNWR 4-6-0 'Claughton', an ex-LYR 4-6-0 and an ex-LYR 4-4-0.

Prior to this decision a few tentative experiments had been made, using either the new initials in full — LM&SR — or in the abbreviated form — LMS — on locomotives retaining their pre-Grouping livery colours. To further confuse things some locomotives appeared carrying their new LMSR numbers but without any evidence of ownership whatsoever. However, once the decision was taken, the application of crimson lake was made with remarkable alacrity, in particular in Scotland, where the new colour scheme seemed to give added character to many of the varied and aged types of locomotive that the LMSR had, in some cases, reluctantly inherited! At this time pre-Grouping loyalties were naturally very strong, but the new livery seems to have been generally well received — despite latter-day stories about the reluctance of Crewe in particular to paint its locomotives in Midland colours — a myth I have already challenged in the Introduction.

Passenger carrying rolling stock, and non-passenger carrying stock that nevertheless ran in passenger trains were given crimson lake livery, generally following the former MR style, and it was not surprising that the goods wagons were basically in MR grey livery too!

The prewar years can be conveniently divided into two main livery styles for locomotives: 1923-1927, 1928-1939, plus one short-lived variant (1936) and two 'special' liveries, for streamlined locomotives. Perhaps the prewar period could conveniently be described as 'the red years', whereas what followed was *black*, in more senses than one!

Although in this series, I try to give a broad specification for each livery period or style, in this case the prewar locomotive livery underwent some changes that require more detailed description, as follows:

Locomotives

Passenger livery; first period, 1923-1927

Ex-MR crimson lake, or 'Midland Red' (LM1/C28) for bodywork (ie boiler barrel, cab side sheets, footplate valances and running plates, cylinder covers, footsteps, tender sidesheets and rear, and tender underframes). Only one pale yellow line was applied to the boiler, on the band between the smokebox and the barrel. All lining was in pale yellow, with black edging and this was a shade of yellow chosen to match the gold (gilt) transfers edged with a fine white line and shaded black, which were used for the large numerals. This yellow lining edged the crimson lake on all areas where it was bordered by black, except on cylinder covers where it was not always applied. Many locomotives appear to have had a yellow line on the wheel rims close to the outer ends of the spokes; the wheels themselves being painted black. Black was also used for the smokebox, footplate and splasher tops and most upper surfaces, including the top of the cab (above gutter level) and the tops of tenders and tanks. Buffer beams were in vermilion red with pale yellow lining edged in black.

To begin with, the letters 'LMS' appeared in small serif characters on cab sidesheets or bunker sides, with the large gold (gilt) numerals on the tender, or tankside. These initials were replaced in 1927 by the circular crest. The engine number was also applied in polished numerals on a black painted iron casting on the front of the smokebox, following former MR practice; (this also applied to goods engines).

Goods livery; first period, 1923-1927

The goods engines were painted in unlined black, with vermilion red buffer beams and shanks, without lining. The numerals were in the same gold (gilt) shade, style and location as for the passenger engines, but had only the fine white line (on the right hand and lower edges — as a highlight) to separate them from the black background; being in all probability the same *black-shaded* transfers as were used on the crimson lake engines. A few black locomotives carried the circular LMS crest, and some the initials 'LMS' as on the earliest crimson lake examples, but the standard style was a gold bordered vermilion red panel on the bunkerside or cabside, with the letters 'LMS' in gold, shaded black (see drawing). There were two different styles of panel over the period, with the second version having a more modern appearance. Cast iron numberplates were specified for the top of the smokebox door, as on passenger engines.

Passenger livery; Second period 1928-1939 (except 1936/7)

The main body colour remained the same (LM1/C28) but the number of locomotive types using it was now restricted to the more important and modern.* A major alteration was the use of the letters 'LMS' in shaded serif form on tender and tank sides, instead of the circular crest, and the relocation of the locomotive running number on the cabside of the engine portion of tender locomotives, and on the bunkersides of tank locomotives. The chief consideration was the need to avoid the confusion that had arisen from the 1923 livery style where the *tender* carried the *locomotive* number. Changes of tender sometimes occurred in the course of workshop overhaul, or through an emergency in day to day running, and it was not uncommon to see locomotives carrying one number on the smokebox door numberplate and another, quite different, on the tender. By placing the running numbers on the engines themselves, and simply the initials 'LMS' on the tender, this allowed tender changes to be made without problems (except that sometimes a black tender became attached to a red engine, or vice-versa!). At the same time the practice of placing the number on the smokebox door, as a cast iron plate, was apparently eased somewhat for older types of locomotive, although all *new* locomotives continued to feature it. Tenders now had their own numbering scheme; with a cast rectangular metal plate applied to the rear of the tender superstructure.

The style of application of the crimson lake livery itself remained basically the same, except that the covers of outside cylinders were now definitely specified for yellow lining to the front and back edges. The yellow line sometimes put on the black wheels was not applied (officially) after the mid 1930s. The insignia varied a good deal in this second period livery, and there were serif gold letters shaded black, serif gold letters shaded red, and serif chrome yellow letters shaded red. The pale yellow lining, to imitate gold, was changed to a richer chrome yellow in the mid-1930s. The buffer beam style was not altered, except to receive the richer yellow line. A feature of new locomotives built in the Stanier régime was the bright metal finish given to the valve motion, cylinder ends, buffer heads, wheel rims and bosses and handrails. In certain instances (see illustrations) chromium plate was even specified.

**The locomotives selected were the non-streamlined 4-6-2s, 'Royal Scots', 'Patriots', 'Jubilees', 'Claughtons', ex-LYR (Hughes) 4-6-0s and ex-MR Compound 4-4-0s.*

Mixed Traffic livery; second period 1928-1939 (except 1936/7)

Because the crimson lake livery was henceforth only applied to select passenger classes, 'intermediate passenger' classes (mixed traffic types) were accordingly specified for a new varnished black livery, with single vermilion red lining. The letters 'LMS' and the numbers were applied as just described for the post 1927 passenger locomotives, and cast iron smokebox numberplates were still applied to new locomotives, although some older ones seem to have had them removed, once the numbers were on the engine cabside instead of the tender.

Basically the single red line was applied to the same format as the yellow one on crimson lake engines; there were however variations, as can best be deduced by study of contemporary photographs. The bufferbeams were the same vermilion red, *without* lining; edged black. On new Stanier mixed traffic types the same bright metal finish was given to the valve motion and wheel rims, etc, as has been described above for crimson lake engines.

The lettering and numerals on the lined black locomotives were in vermilion red-shaded gold (gilt) serif characters. To begin with these were in a new style which had the red shading *countershaded* with lake and white to produce a more three-dimensioned effect. (There is little doubt that this version was produced specifically for the lined black locomotives, but it did find its way on to some crimson lake and plain black examples as well.) There were also some with plain gold (gilt) transfers (actually *black* shaded!) and finally in the later 1930s a chrome yellow version was favoured, with vermilion red shading.

Goods livery; second period, 1928-1939 (except 1936/7)

Virtually no change, except that the all-black livery had bright metal motion parts etc, on newly-built Stanier locomotives. The revised arrangement for cabside numerals and tender lettering was adopted. In the case of tank engines, the numerals were now normally on the bunkersides and the letters 'LMS' on the tanksides. Plain vermilion red was used for bufferbeams and shanks. The colour of the insignia varied, as with the lined black mixed traffic locomotives, being in turn gold (gilt) shaded red or plain gold, then chrome yellow shaded red, or in some instances plain yellow. (Probably hand-painted retouching produced this latter type.)

The 1936/7 livery Variant

This was a shortlived attempt to change the style of the lettering and numerals, on all types of locomotive, to a more up-to-date sans-serif form (somewhat similar to the LNER 'Gill sans' style). First of all a 'Compound' 4-4-0 No 1099 was repainted in November 1935 with plain gold sans-serif insignia; then No 1094 of the same class was given a black shaded version, but what actually appeared on most new locomotives, and some repainted ones, during 1936/7, was in gold shaded in vermilion red (see illustrations). The locomotive colour schemes were not altered, ie crimson lake engines remained crimson lake, etc, but there were a few goods engines finished in plain black with plain yellow (or perhaps gold) insignia — most however, had the red shading. The cast iron smokebox numberplates were altered to the new style numerals, and being very durable they continued in use until the last days of the LMSR in some cases, whereas the transfers (or perhaps hand-painted) insignia reverted to the earlier serif style from late 1937, early 1938 onwards.

The special streamlined liveries; 1937/1939

The 'Coronation Scot' express train of 1937 for which the first of William Stanier's classic new 'Princess Coronation' Pacifics were specially built in streamlined form, featured a livery of Caledonian blue (Carter calls it Royal blue; his chart reference is C24) which was applied overall to locomotive and tender, and the carriage sides. A darker, Prussian or perhaps Navy blue, was used for the locomotive driving wheels and as a background colour to the chromium plated nameplates, and as lining to the edges of the silver (aluminium paint) broad bands which ran horizontally along the length of the locomotive and train, commencing in a vee-shape on the front of the engine. The style of insignia was basically the 1936 sans-serif, but without shading. Only the first five locomotives, Nos 6220-6224 received the blue livery.

When five further streamlined 'Pacifics' were delivered the livery was changed to the standard crimson lake and with horizontal gold bands instead of silver, edged with a fine vermilion line and a bolder black border; still using the sans-serif insignia. This was the intended livery for the new 1939 'Coronation Scot' train, which was exhibited in America (complete with No 6229 *Duchess of Hamilton* masquerading as No 6220 *Coronation*) and which was stranded there by the outbreak of World War 2; of which more anon.

Passenger Stock

Unlike the locomotive stock, where economies reduced the number of types carrying crimson lake livery, the passenger carriages retained this colour scheme throughout the prewar years; only the style of lining and some insignia detailing was changed.

The broad specification was as follows:

From 1923 until late 1933/early 1934

Crimson lake (LM1/C28) bodysides and ends. Gold (corridor stock) or pale yellow (non corridor stock) lining each side of a black band on the raised beadings of wooden-panelled stock; a fine vermilion red line separating the crimson lake. The same livery was applied to flush-sided steel-panelled stock; giving an impression of panelling.
Black underframes and bogies
Black roof between rainstrip and cantrail; grey above rainstrip.
Letters 'LMS' and descriptions, ie 'Sleeping Car' in 4in high elongated serif letters in gold, with countershaded red to the left and black to the right. It is interesting to note that the red shading for the lettering on carriages was on the *opposite* side of the letterform to that applied to locomotives, and had an additional *black countershading* which was not used on locomotives at any time.

From 1934 until 1939

Crimson lake (LM1/C28) for bodysides; the ends also until 1936, when they became black.

Simplified lining with two single yellow lines between cantrail and windows and a single black band edged each side with a yellow line just below the windows. The yellow being a darker chrome, on corridor and non-corridor stock.
Black underframes and bogies, and ends after 1936. Roof in aluminium metallic finish; or grey.

Running numbers in sans serif shaded numerals; other insignia unchanged.

The 'Coronation Scot' trains, 1937/39

The 1937 train was finished to match the streamlined Pacifics, in Caledonian (or Royal) blue (C24). Sans serif silver insignia with a dark blue outline was used, but the train name was carried in black on white on the carriage roofboards. The 1939 train was in crimson lake and gold, this time with the train name painted in gold outlined in black above the windows. (See the locomotive descriptions for further details of the lining, etc which was carried along the carriage sides to match the locomotive.)

Railcars and multiple-unit electrics

Normally these were in the standard crimson lake, but sometimes lined in special fashion (see photo of the Leyland railbus, for example). Electric stock followed the styles adopted for non-corridor locomotive-hauled carriages. The exception was the diesel three-coach articulated railcar set introduced in 1939, which had a bright red and cream livery, with silver roof.

Other Stock

Passenger brake vans, milk vans, fish vans, horse boxes and other specialised non-passenger carrying stock

All stock that was permitted to run in passenger train consists, received full crimson lake livery from 1923 to 1933/4, complete with pale yellow lining-out. Then when the simplified passenger carriage livery was introduced, all these vehicles (except passenger brake vans) received plain crimson lake, devoid of lining but retaining the shaded serif letters for the LMS insignia, and for special descriptions such as 'Milk Van'. The numerals were in unshaded sans serif style.

Goods wagons; from 1923-1936/7

Light to mid grey, bodywork sides and ends, and solebars and headstocks.
Black running gear and buffer heads and shanks and all details below solebar.
White unshaded sans serif numerals and lettering.

Goods wagons; from 1936/7-1939

Bauxite brown (or red) bodywork sides and ends, solebars and headstocks and sometimes roof; otherwise aluminium or grey was used for the roof.
Black running gear and buffer heads and shanks and all details below solebar.
White unshaded sans serif numerals and lettering, all details being in small characters grouped at the lower left-hand end of the bodyside.

Service vehicles

Loco coal wagons, some crane match wagons and the plough brake vehicles used in engineers' trains appear to have followed much the same style as revenue-earning wagons except for the insignia carried. However ballast and sleeper wagons were in red oxide and some crane match wagons were painted black.

Containers

Open type containers carried the wagon livery; ie grey from 1923-1936/7 and bauxite afterwards. Ordinary covered containers were in crimson lake with yellow lettering, and sometimes the LMSR crest was applied. For insulated types (such as for meat traffic) there was a special white livery with black lettering, etc. Slogans were carried for publicity purposes on many types; in particular furniture containers.

10

10
Looking quite splendid, groomed to haul the Royal Train, 'Claughton' class 4-6-0 No 5944 was photographed on Bushey water troughs on 15 August 1925, with the 11.00am Royal special from Euston. The locomotive is in the new crimson lake livery, complete with numberplate on the smokebox door, and the cab sidesheet carries the initials LMS, a short-lived idea prior to using the new coat of arms. Until World War 2 the Royal Train continued to carry the livery of the former LNWR, known unofficially as 'plum and spilt milk', at the specific request of the late King George V, who apparently preferred it to the LMSR crimson lake scheme.
F. R. Hebron/Rail Archive Stephenson

11
The Midland Railway-inspired crimson lake livery with pale yellow and black lining and edging, sat very happily upon the standard 4-4-0 'Compounds', which were to carry Derby practice to most corners of the LMSR system; becoming known as the 'Crimson Ramblers' as a result. No 1115 depicts the standard livery for passenger locomotives of the 1923-1928 period, with the LMSR coat of arms, or crest, upon the cabside and the locomotive number on the tender; plus a MR style cast numberplate on the smokebox door. *LPC*

12
The locomotives of the Scottish constituents of the LMSR seemed to carry the crimson lake livery exceptionally well. It certainly suited the ex-HR 'Clan' class 4-6-0s, as No 14768 *Clan Mackenzie* shows, in this official broadside view. In the first version standard crimson lake livery from 1923-1928 the numerals were on the tender, or on the tank sides, and the locomotive carried the coat-of-arms on the cab sides. The lettering was in gold leaf with black shading, including the name whenever (as in this case) it was applied hand-painted, and not as a cast plate. This style of lettering was exactly as used previously by the MR. *Ian Allan Library*

11

12

13
Fowler's 'Royal Scot' class 4-6-0, in its original form was very obviously Derby-inspired in appearance, and the livery emphasised this! Features to note on the 1923-1928 crimson lake engines are the MR style numerals and the lack of lining for boiler bands except at the edges adjacent to the smokebox and cab front; the pale yellow lining out on footsteps on engine and tender; lining out on the black-painted wheels immediately adjacent to the outer ends of the spokes, and the lining with black edging to the outer edges of the bright vermilion bufferbeam and buffer shanks. On some classes including the 'Scots' there was a band of crimson lake above the vermilion area; as seen here on No 6110 *Grenadier Guardsman*, photographed at Camden. *LPC*

13

14
Goods engines in the 1923-1928 period were painted plain black, with gold leaf numerals and vermilion bufferbeams; without lining. On the cabsides a panel of vermilion carried the initials LMS, with the lettering and edging in gold leaf, shaded black. Two versions of this panel were produced (see drawing 16) and the first version is seen here on ex CR McIntosh '179' class 4-6-0 goods engine No 17909. *Ian Allan Library*

14

15

15
Standard Fowler Class 3 0-6-0T No 16624, photographed at Forres on 16 May 1928, when newly delivered from Beardmore's. Plain black livery with gold leaf numerals and second version of the 'LMS' vermilion and gold panel with rounded corners, on the bunkerside. *H. C. Casserley*

16
Left: The elegant numeral style of the 1923-1928 livery, based upon MR practice, in gold leaf with black shading; plain gold leaf on black engines. Two sizes of numerals were used, 18in and 14in high. The 14in style differed in being slightly more condensed.
Right: Lettering style on black engines for the cabside, or bunkerside, panels. The first version was replaced by the second in 1926. Background colour was vermilion, and the letters and border were in gold leaf, shaded black.

16

12345 18"
67890

32" 28" 11" 7" LMS 2" RAD 2"
11" LMS 2" RAD

17

17
In 1928 a revised standard livery was introduced, retaining the same livery colours, but placing the locomotive number on the cabside of tender engines and replacing the coat-of-arms, or crest, with the initials LMS on the tender. 'Royal Scot' class 4-6-0 No 6149 *Lady of the Lake* (later renamed *The Middlesex Regiment*) displays the revised scheme to perfection in this photograph taken at Crewe North. The large cabside numerals were shaded black and were in the traditional Derby style (see drawing) and were 14in high. No lining-out to boiler band except for a single yellow line adjacent to the smokebox. No lining on the wheels. The crimson lake was carried up to the gutter line above the cab; above the roof was black. The small brass oval below the nameplate depicted the LNWR locomotive of the same name. *T. G. Hepburn/Rail Archive Stephenson*

18

18
In the revised livery, tank engines carried the initials LMS on the tanksides, and the numerals on the bunkersides. Crimson lake was still applied at first to some selected passenger tank engines, as seen here on the first of the Fowler Class 4P 2-6-4Ts, No 2300, photographed at St Albans on a St Pancras suburban working. It was however not long before black with red lining was substituted for the crimson lake, but over half of the first 25 engines of this class were delivered in crimson lake. One example, No 2313 carried the name *The Prince* painted above the initials on the tanksides, from 1928-1933, to celebrate the visit of the Prince of Wales (later the Duke of Windsor) to Derby Works in February 1928, when the engine was nearing completion. Note the standard cast iron numberplate fitted to the smokebox door, and the shallow vermilion red buffer beam area with a band of crimson lake above it. The ex MR 0-4-4T in the righthand background is still in the first version of the crimson lake passenger livery for tank engines, with the LMSR crest on the bunkerside. No 2300 has the small 10in high numerals, shaded black. *LPC*

19

In 1933 the express passenger engine, 'Royal Scot' class 4-6-0 No 6100 *Royal Scot* was sent to North America by the LMSR for exhibition purposes. For this it was fitted with an electric headlight and warning bell, and the smokebox numberplate was replaced by a polished metal plate carrying the name of the train 'The Royal Scot', (*not* the name of the locomotive, which was simply *Royal Scot* on the nameplates). Although specially finished for this event, the burnished steel buffer heads, and the outer ring and hinges to the smokebox door were a feature often found on other examples of this class running on the LMSR (and on the smaller 'Patriots' and Compound 4-4-0s). The shallow vermilion red area on the buffer beam is well depicted, with the yellow and black surround; above it was painted crimson lake. Smoke deflectors added to the front end. The headlight was removed when the locomotive returned to Britain, but the nameplate and bell were retained as a souvenir of the visit. *British Rail*

 19

20 

20

When William Stanier, (later Sir) took over LMSR locomotive affairs he did not alter the existing livery schemes (unlike many CMEs in the past) and the crimson lake scheme suited his new express passenger locomotives admirably; as portrayed by the superb 'Princess Royal' class Pacific No 6201 *Princess Elizabeth* in this official broadside of the locomotive when new. Original Derby-style tender, with an additional lined-out crimson lake panel at the footplate end. Polished wheel bosses and rims, valve motion and spring details, and cylinder end covers. *LPC*

21

In original condition, with domeless boiler, Stanier's Class 5XP 'Jubilees' were extremely elegant locomotives, with polished wheel bosses and rims, motion, and metal fittings including the handrails, and reversing lever. The cylinder end covers were in bright metal, and these details gave the crimson lake livery the finishing touch. No 5564 is illustrated, brand new and not yet carrying the name *New South Wales*. Note the low placing of the 12in cabside numerals, in line at their base with the lining on the running plate above the driving wheels; whereas the initials LMS were carried centrally on the tender sides. *British Rail*

21 

22
In honour of the Silver Jubilee of King George V and Queen Mary in 1935, the 'Jubilee' Class 5XP 4-6-0 No 5642 was taken into works and given a special black and chromium plate finish, at the same time exchanging name and number with the existing No 5552 *Silver Jubilee*, which became No 5642 *Boscawen*. The new No 5552 had a very glossy black paint finish and the numerals and lettering were in polished relief metal, in bold 'Grotesque' sans-serif style. The smokebox numberplate was also in a particularly neat sans serif style; unique to this locomotive. All the boiler bands, and the top feed cover (again in a style unique to this locomotive), the outside steampipes, the entire cylinder ends, the handrails and sundry other external fittings, were given a chromium-plated finish. Before going into normal revenue-earning service, the locomotive went on an exhibition tour of the LMSR system. Photographed at Camden shed. *P. Ransome-Wallis*

23
Fowler 'Royal Scot' class 4-6-0 No 6160 *Queen Victoria's Rifleman* with Stanier tender and smoke-deflectors added, in correct early to mid-1930s crimson lake livery, with black shading to the 14in numerals, and the tender lettering. The power classification, 6P, is placed above the numerals on the cabside; previously it had been higher, between the cab window and the rear cutaway. Comparison with photo 17 shows the difference in appearance created by the larger Stanier tender and the smoke deflectors. *British Rail*

22

23

24
LMS S
12345 5
67890

24
The lettering and numerals of the revised livery, introduced in 1927 and used until 1947 (with exceptions, as described later). There were three sizes of numerals: the existing 14in plus new style 12in and 10in. The 14in letters for 'LMS' were applied in gold leaf shaded black; in gold leaf with vermilion red shading, or with blended red to white countershading; in plain gold, plain yellow and (in latter days) plain yellow with red shading. There were similar colour variations for the numerals.

25

The nameplate of 'Royal Scot' class 4-6-0 No 6118 *Royal Welch Fusilier*, with cast gun-metal regimental crest carried below. This picture shows well the bold lining style applied to the crimson late engines. The $\frac{1}{2}$in yellow line was pale (straw) until 1935 when it was changed to a richer yellow, more like 'old gold'. Nameplates were normally in polished brass, with a black background. *Ian Allan Library*

26

Evidently someone on the LMSR, in authority, wanted a more 'modern' style of lettering and numerals for locomotives, and in 1936 (March, to be precise) a 4-4-0 'Compound' No 1094 was repainted with bold sans serif numerals in yellow shaded black. This cabside detail also shows the style of lining-out for locomotives with Derby-style cabs, and power classification '4P'. *British Rail*

26

25

27

28

L M S

12345

67890

27

The sans serif style began to appear on some repainted and some new locomotives, but the black shading was replaced by vermilion red and the yellow by gold leaf. 'Patriot' Class 5XP 4-6-0 No 5502 *Royal Naval Division* demonstrates the style, sometimes referred to as the '1936 style'. *British Rail*

28

The '1936 style' sans serif alphabet. The letters 'LMS' were in 14in high characters and the numerals were 10in high. Smokebox numberplates were also produced to this style.

29

The magnificent streamlined 'Princess Coronation' Pacific, introduced by Stanier in 1937 specifically to haul the new 'Coronation Scot' train, were strikingly finished. The basic livery was intended to be a match of the Prussian blue of the former Caledonian Railway (and appears to have been an accurate one). To this was added silver for the lettering and numerals and for the bold stripes, or 'cheat lines' which commenced at a point low down on the nose, just above the coupling, and then widened to run along the entire side of the locomotive and tender; being continued along the carriage sides of the train itself. The numerals and lettering were basically to the '1936 style', but without shading. No 6222 *Queen Mary* is illustrated. External metal details, such as handrails, were given a chromium-plated finish. Photographed in August 1937. *British Rail*

29

30

30

Left: The nameplate of the first of the 'Princess Coronation' class, No 6220 *Coronation*, which was chromium-plated and had a dark blue background. The same dark blue (navy blue) was used as a fine $\frac{1}{4}$in lining to each side of the silver stripes (actually aluminium paint was used). Note the chromium finish to the handrail at the top of the picture, and to the sandbox filler cap, and the Crown in polished relief metal finish

Below: Drawing of 'Princess Coronation' nameplate.

British Rail

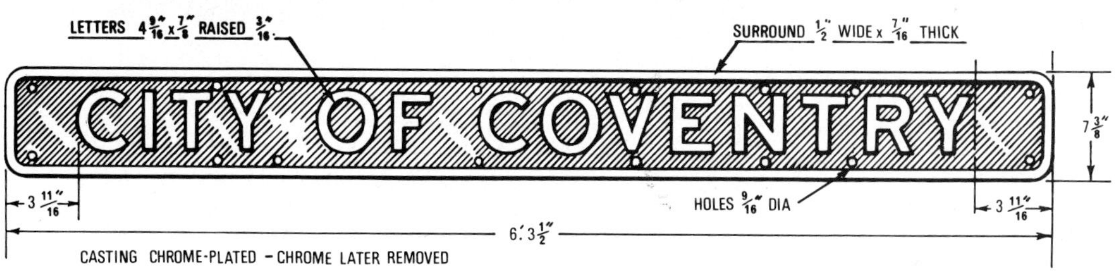

31
In 1935 Stanier had rebuilt the experimental Fowler high-pressure 4-6-0 No 6399 *Fury* into a taper-boilered version of the 'Royal Scot' class, renumbered No 6170 and named *British Legion*. This was the only taper-boilered 'Royal Scot' to carry a single chimney (later replaced by a double), and the only one to carry LMSR crimson lake livery; all subsequent rebuilds appearing in black, as described later. No 6170 was a most handsome machine, as can be seen in this picture of the locomotive heading the up 'Mancunian' passing Crewe. Shallow vermilion red panel to the bufferbeam, with crimson lake band above. *Real Photos*

31

32
A special high-quality crimson lake livery finish was bestowed by Stanier upon his five superb non-streamlined 'Princess Coronation' (or 'Duchess') class 4-6-2s Nos 6230-6234. Here No 6231 *Duchess of Atholl* is seen on shed, with the chromium-plated handrails to cab and tender well depicted. The lining-out was gold, instead of yellow, with a fine red line to each side of it, and the 12in numerals and 14in lettering reverted to the pre-1936 pattern, but in gold leaf shaded with vermilion. The nameplates were in polished chrome with black background, and certain other small external fittings (such as sandbox filler caps) were also chromium-plated. *P. Ransome-Wallis*

32

33

34

35

33
The abandonment of the '1936-style' sans serif characters (except on the streamlined-engines) was decided upon only some two years after their introduction! In their place, the 1927-style characters were again used, but the gold finish was replaced by yellow (except on Nos 6230-6234) and vermilion red shading replaced the black, on many repainted engines. These bright and contrasting colours had the effect of making the crimson lake itself look darker in tone. 'Royal Scot' class 4-6-0 No 6139 *The Welch Regiment* was photographed at Crewe in 1939. Note the blank backing-plate above the nameplate, ready for the cast gun-metal regimental crest; not fitted.

34
Only the first five streamlined 'Princess Coronation' Pacifics were given the Caledonian blue livery, and then the standard crimson lake was substituted. The bold stripes were applied in gold paint with a fine $\frac{1}{8}$in vermilion line to each side, with a black line (somewhat thicker) separating it from the crimson lake. Lettering and numerals were in gold to match; external metal details were still chromium plated, including the nameplates which had a black background. In 1939 a new 'Coronation Scot' train was constructed, and prior to use in this country it was sent on an American Tour (see also photo 50). The locomotive and train were both finished in the crimson lake and gold livery. No 6220 *Coronation* (in reality No 6229 *Duchess of Hamilton*, which exchanged identities to become No 6220 for the duration of the Tour, being a newer locomotive). is seen here on the Baltimore & Ohio RR, complete with headlight and bell, to conform with American Railroad regulations.
Ian Allan Library

35
Cabside detail of No 6220 (actually No 6229) in the crimson lake and gold livery scheme with '1936-style' numerals, but without shading and outlined in black. Note the power classification '7P' placed *below* the numbers. Fireman J. Carswell and Driver F. C. Bishop, both seen here, were the men that took the locomotive to America, but Bishop fell ill and R. A. Riddles (later destined to be the last cme of BR steam) did most of the driving on the Tour.
Crown Copyright, National Railway Museum, York

36

36
Former Caledonian Railway mixed traffic 4-6-0 No 14800 (ex No 956), photographed in May 1929 in lined black, following the decision to restrict the use of the crimson lake, circa 1928. The new livery for mixed traffic locomotives (or as they were sometimes called 'intermediate passenger locomotives') was black with red lining. This replaced the crimson lake at first applied by the LMSR to many pre-Grouping types in this category.

37
New mixed traffic locomotives were delivered in the lined black livery, and foremost amongst these were the classic 'Black Fives' (or 'Black Staniers') which of course gained their nickname from this livery, thereby segregating them from the outwardly very similar Stanier 'Jubilees', which were in crimson lake. Class 5 4-6-0 No 5157 *The Glasgow Highlander* (one of only four of the class named in LMSR days) displays the lined black livery to perfection, and shows the gold lettering and numerals with red countershading that were introduced for this livery. 5P5F power classification on cabside below windows, and tablet catcher on cabside adjacent to the figure 7, which it partially obscures. Note the polished steel cylinder end covers, wheel rims and bosses and valve motion.
Crown Copyright, National Railway Museum, York

38
Stanier Class 4P 2-6-4T No 2629 in lined black livery; photographed in September 1938. The smokebox door numberplate is in the '1936-style' sans serif numerals whereas the tank and bunkerside carry the yellow, shaded red, characters which replaced the sans serif style in that year. The vermilion red bufferbeam has a narrow black border around it. No lining on cylinder covers or along the running plate. *British Rail*

37

38

39

40

39
The prewar film has failed to render the red lining in this photograph of Fowler 2-6-4T Class 4P No 2341, but close inspection of the original print shows it to be there (but faded). It is included here to show the use of unshaded insignia on a black engine, sometimes resulting from the use of *black* shaded transfers, presumably done for expediency in the workshops!
P. Ransome-Wallis

40
The goods engine livery was unlined black, with vermilion red bufferbeam and shanks, throughout from 1923-1947; only the styles and location of insignia varied, and these changes followed the same pattern as those made to the passenger and mixed traffic types. The lettering and numerals were in unshaded gold to begin with, but from about 1929 onwards red shading was sometimes used. This Beyer-Garratt, No 4994, photographed north of Elstree in a down empty wagon train, carries the plain gold version. The letters LMS were closely spaced to fit the cabside on this particular class.
F. R. Hebron/Rail Archive Stephenson

41
The '1936-style' insignia was applied *unshaded* to a few of the new Stanier 2-8-0s then being built by Vulcan Foundry in the period 1936/37, as seen here on No 8042 in glossy black livery, with polished metal wheel rims, bosses and motion. These were an exception, however, because the majority of locomotives which received this style of insignia had them in gold shaded red.
British Rail

41

42

42
The '1936-style' was also applied to smokebox numberplates, and examples of these lasted considerably longer in service (even after the locomotive had been repainted and had lost the sans serif on the cabsides, etc). These numberplates were very legible, as is demonstrated by Stanier 8F 2-8-0 No 8207, photographed pounding along on an up goods train near Elstree.
E. R. Wethersett

43

43
Although the design of this Class OF 0-4-0ST has sometimes been ascribed to Stanier, it is doubtful if he had much to do with them; being basically a Kitson-type industrial design, introduced in 1932. No 1540 displays the plain gold insignia, and small 10in numerals on the plain black livery. *LPC*

44

44
A non-standard, but very attractive livery was bestowed by the makers upon Sentinel 0-4-0T shunter No 7164. The engine is in glossy black with a gold line and a thinner red one inside it. The lettering and numerals are in gold, boldly countershaded to give a relief effect. A most attractive and simple machine for the modelmaker! *LPC*

LMSR crimson lake (LM1/C28).

'Princess Coronation' 4-6-2 No 6224 *Princess Alexandra* in blue and silver livery, at Shrewsbury in 1938. Taking water whilst on a 'running-in' turn from Crewe. *Colour-Rail/P. B. Whitehouse*

45
From 1923 until the end of 1933/early 1934 all the LMSR passenger carriages were painted in what was virtually the style of the former Midland Railway, with fully lined crimson lake livery; only the crest and insignia being changed to suit the new company. The crimson lake was also applied to the carriage ends, but without the elaborate black and gold, or yellow lining with vermilion edging. This lining was on the beading of the panelwork (stock being basically of wooden construction above the underframes, with mahogany or steel panelling) and the effect was to emphasise the panelwork, in a most attractive way. The LMS crest was carried on the lower bodyside with the initials above in gold, shaded left in graduated lake, or red and countershaded right in black. The roof was black between rainstrip and cantrail and grey above the rainstrip. Bogies, underframes and all detail below the body, buffers and end beading and fittings were all black. Note the elaborate etched-glass toilet window in the centre of this elegant semi-open first class carriage No 15933 of circa 1930. *Ian Allan Library*

45

46

47

46
The LMSR introduced steel panelling for its carriage construction during the 1920s, and also purchased a quantity of all-steel carriages from outside builders. To begin with the livery style of the wooden panelled stock was retained and 'pseudo' panelling was painted upon the steel sides! Illustrated is an all-steel brake third open, No 7670 built by the Leeds Forge Co in 1926. Note the riveted steel roof. *British Rail*

47
All-steel third open built by Cammell/Laird in 1925/26, with revised livery style, simplified to avoid vertical waist and eaves panel divisions; giving a more modern appearance. *British Rail*

48

48
A Stanier design combined kitchen and third class dining car, on six-wheel bogies, in steel-panelled form with flush surfaces but retaining traditional lining-out to simulate beading. Roof finished in aluminium/silver paint. No 102 also shows the attractive style of lettering used for the words 'Dining Car' and the sans serif numerals for the running numbers. *Ian Allan Library*

49

49
Third class open of 1938 Stanier-build, with the revised and simplified style adopted, devoid of panelling and certainly more suitable for modern flush-surfaced rolling stock. Two fine yellow (instead of gold) lines above window level and a black line edged with yellow lines along the waist. Grey roof and black carriage ends. *Ian Allan Library*

50
Top: The previous year, 1937, witnessed the introduction of the 'Coronation Scot' express train, between London and Glasgow. For this a special livery of Caledonian blue and silver was applied (see photo 29) and the lettering and numerals were in silver sans-serif. Illustrated is Kitchen car No 30089 of the 1937 train.

Bottom: In 1939 Stanier introduced a completely new train set for the 'Coronation Scot' service. This purpose-built set was finished in crimson lake and gold livery (see also photo 34) and the horizontal gold stripes were carried round to the rear of the end brake vehicle, as seen here. This brake was part of an articulated twin with first class accommodation, Nos 56000/1. The brake end had no gangway vestibule connection. The name 'The Coronation Scot' was painted in gold sans serif lettering on the carriage sides above the windows, whereas the 1937 train had carried normal black on white roofboards. Note the flush steel valances carried down between the bogies. This train was sent to North America for a Tour in 1939, where it was marooned by the outbreak of World War 2, and it spent the entire war years there in use as an officers' mess in a military camp! It never ran as a complete luxury passenger-carrying set in Britain, because the LMSR decided not to reintroduce the train in the difficult immediate postwar period. Note the oval buffers fitted to this train.
Crown Copyright, National Railway Museum, York

50

The non-streamlined version of the 'Princess Coronation' class (the 'Duchess') had a de-luxe finish to their crimson lake livery; illustrated is No 6232 *Duchess of Montrose* on a Crewe local at Shrewsbury in 1938. The locomotive is in original condition with single chimney, and before the addition of smoke deflectors.
Colour-Rail/P. B. Whitehouse

Stanier Class 2P 2-6-2T No 91 in ex-works condition in the lined black 'intermediate passenger livery', with red shaded gilt numerals and lettering. Photographed at Derby in July 1938. *Colour-Rail*

Class 4P 'Compound' 4-4-0 No 1111 in standard 1930s crimson lake livery; featuring 'countershaded' insignia. *Steam & Sail*

A wartime photograph taken on the Cromford and High Peak line, showing ex NLR 0-6-0T No 27527 at Middleton in September 1943. Note the closely spaced plain gilt letters and numerals! Plain black goods engine livery, with freshly painted smokebox. *Colour-Rail*

51

52

51
The livery for non-corridor passenger stock followed the same style as the corridor stock and the use of 'pseudo-panelling' was a feature of early steel panelled construction, being emphasised even more by the more numerous doors and windows! This attractive view of 'Push and Pull' brake third No 24404 also shows well the 'LMS' insignia. Yellow was used instead of gold for the lining-out on non-corridor stock, but otherwise the application was the same and included plain crimson lake ends (until 1936, when only 'Push and Pull' carriages retained the crimson lake ends, as seen here, all others being painted black). A small destination board is fitted above the windows near the centre of the carriage; in this case reading 'Pye Bridge' in black lettering on white. The locomotive is former MR Class 1P 0-4-4T No 1340, in black livery with red lining. *British Rail*

53

52
Sentinel-Cammell steam rail motor-car No 4151 built 1926/27, with third class accommodation only. The engine unit, nearest the camera, is articulated with the passenger saloon, which had 44 seats. One of 14 steam rail-motors that the LMSR operated for about 10 years. The crimson lake livery had the beading picked-out in yellow and black, including the driving ends.
Ian Allan Library

53
Electric rolling stock followed the standard livery pattern, and the same 'pseudo-panelling' was painted on the flush steel sides to begin with, later being simplified. The driving ends were plain crimson lake by way of contrast and the bufferbeam and shanks were in bright vermilion red. Note that there is no LMSR crest on the sides of this motor brake third No 8881, built in 1927.
Modern Transport

54
The electric stock constructed during the Stanier period was considerably more up-to-date in concept, and it looked extremely smart in the simplified crimson lake livery, with sans serif numerals and black and yellow waist lining and the LMSR crest. No 29287 is the leading car of a three-car set on a Liverpool Central-West Kirby working, photographed at Birkenhead North. Earlier types of electric rolling stock also received the simplified livery in the late 1930s. *W. Hubert Foster*

54

55

55
Leyland diesel railcar No 29950, one of three four-wheeled vehicles delivered in 1934, and basically of bus-type construction. These 40 seat third class only railcars were in crimson lake livery, complete with the LMSR crest on the centre doors. No accurate record has come to light concerning the colour of the two light bands around the waist and skirt, but it is assumed that they were either pale grey or cream. No lining-out was applied, and these vehicles presented a very modern appearance for their time. *LPC*

56
A remarkable train produced by Stanier was the three-car diesel set, introduced in 1939 and unfortunately overtaken by the outbreak of war, which did not favour such experiments. It had an articulated layout and fully streamlined nose ends (reminiscent of German high-speed railcars) and the livery was bright red and cream with black waist line and thin black lining to the upper band of red, with silver roof. Lettering and numerals were in sans serif cream, and no LMSR crest was applied. Nos 80000/1/2 (the latter is nearest in this picture) were stored during the war years and afterwards rebuilt as a maintenance unit for the MSJA electric line; being reduced to two cars in the process. A sad end to a fine experiment! *LPC*

56

Ex CR 4-6-0 No 14622 standing at Oban shed in 1938. 'Intermediate' lined black livery and gilt insignia with vermilion red shading. Note how the red lining has all-but disappeared on parts of the engine (splashers and running plate in particular). *Colour-Rail*

'Cauliflower' goods 0-6-0 No 8592 standing by the coal stage at Bletchley in June 1938; plain gilt lettering, with white outline evident. *Colour-Rail/L. Hanson*

57
Non-passenger carrying rolling-stock which nonetheless was hauled attached to passenger carriages was accorded full crimson lake livery, complete with yellow and black, or plain yellow lining, until the introduction of the simplified livery scheme for passenger carriages, when all lining-out was abandoned (see photo 59). On modern steel panelled stock, such as this early Stanier (1932) designed six-wheeled passenger brake van, No 2860, the 'pseudo-panelling' was applied to match the rest of the train. The legend 'To carry 6 tons' was painted in yellow italic script on the lower right hand corner, below the number and a grey panel was painted above (at waist level) for chalked inscriptions.
Real Photos collection

57

58
Wooden-bodied stock, such as horse boxes and fish vans, which could run in passenger trains, received the crimson lake livery, but the style of application of the yellow and black lining depended upon the form of construction. On this fish van, No 7674, only yellow lining was applied and this was restricted in use to the edges of the raised framing. *Real Photos collection*

58

59
When Stanier introduced the simplified lining for passenger carriages, the non-passenger stock in the lined crimson lake livery (except passenger brake vans) was henceforth painted without any lining-out, and with sans serif unshaded numerals. The 'LMS' insignia however, and special inscriptions such as 'Insulated Milk Van', as seen here, retained the shaded serif characters.
No 38551 was a new-type insulated milk van designed to carry milk in churns, and was built at Derby works in 1935. *Modern Transport*

59

60
The LMSR wagon livery was basically a perpetuation of MR practice, with a grey finish for bodywork and solebars, with white lettering and numerals, and black for running gear, buffers and all details below the solebars. The exact shade of grey appears to have varied somewhat, but newly painted stock (see next photo) was in a light shade. This delightful picture shows tractors being loaded on to a special train composed of medium open goods wagons. *Ian Allan Library*

61
A publicity pose for a complete train of cattle wagons and brake van, all in fresh light grey livery. The brake van has the number in white upon a black panel; all other lettering and numerals being in white on grey. Note that the solebars are all grey and also the rear bufferbeam of the van. An ex MR 0-6-0 goods engine is just visible at the head of the train, which was photographed in 1923. The cattle wagons carry the letters LMS to the left of the centre doors, and the inscription 'Large' to the right.
Crown Copyright, National Railway Museum, York

60

61

62

62
The size of the letters 'LMS' varied considerably, and being hand-painted it also varied slightly in style (particularly the drawing of the letter 'S') and in thickness. This 'Bulk grain' hopper wagon, photographed in grey livery in 1936, has quite small and 'bold' letters. Note the patch of black paint on the solebar, to carry the oiling date and district number.
Crown Copyright, National Railway Museum, York

63

64

63
Also photographed in grey livery in 1936, this 20ton brake van makes an interesting comparison with the one depicted in photo 61. In particular the change of style and location for the running number. Note that the handrails were all painted white. The roof on covered vans was grey, but aluminium was specified in later years (1936).
Crown Copyright, National Railway Museum, York

64
An insulated meat container, in white livery with glossy black lettering, detail and roof. The colour of containers was grey for open containers (changed to bauxite after 1936) and for covered containers either crimson lake see photo 101, or white (for insulated types). The flat wagon is in grey livery. *Ian Allan Library*

65
In 1936/37 the livery for goods wagons was changed from grey to bauxite brown (or red, being an indecisive shade between red and brown). At the same time the use of the large letters LMS was abandoned and these appeared at the left hand end of the body in small characters, painted white together with the running number and other details. This 12T shock absorbing wagon is in the bauxite livery (including solebars and headstocks and buffer shanks) and has three vertical white stripes to denote its special type; note the exposed control springs on the underframe.
Crown Copyright, National Railway Museum, York

65

2: War and Postwar 1939-1947

I have already discussed the effects of the war years upon the LMSR in the Introduction, and the photographs that follow tell their own story. Basically it was a period of 'make-do and mend'. The crimson lake livery was abandoned for locomotives and replaced by plain black, either with the existing insignia retouched, or in new transfers of yellow, shaded red. The carriages received plain maroon if repainted, devoid of lining; and I have expressed my belief that this maroon was the result of wartime shortages both of top quality paint and of manhours to apply it. This state of affairs was most evident in the case of goods wagons, which scarcely saw a paint brush! Wartime wooden-bodied wagons had bare planks, with just patches of bauxite where white lettering was needed. If there was a standard livery for the LMSR when the war ended it could perhaps be summed-up as overall grime with glimpses of black and maroon!

In the winter of 1945/6 the LMSR directors evidently began to consider the livery to be applied for their weary postwar railway system. Quite apart from an appalling backlog of repairs and maintenance, they were faced with chronic staff shortages — personages such as engine and carriage cleaners were virtually non-existent because the normally dirty environment of many depots had been dramatically worsened by bomb-damage and lack of repair and they were *not* inviting places for work — hence staff-recruitment was a major problem. Obviously it was pointless to consider a return to the red and gold of the final prewar streamliners, if no-one cleaned it!

A large drawing was first of all produced (fortunately it has survived and is now in the National Railway Museum's collection), showing the side elevation of a non-streamlined 'Princess Coronation' Pacific. This was to portray the proposed livery for express passenger engines, and the basic colour was a dark blue/grey similar to that used by the RAF for their road vehicles. This blue/grey was embellished with wide maroon boiler bands, maroon and straw yellow lining out and straw yellow, (to match gold) insignia. To assess the likely effect of the proposed lining and insignia Pacific No 6235 was decked-out in photographic grey (without the boiler bands painted) at Crewe, when de-streamlined. Evidently there was then some internal dispute, with some directors preferring a return to the 'Midland Red' — or more accurately, maroon — and the next stage was the granting of authority to paint three locomotives in March 1946 in full experimental livery schemes, for comparison. Details of these were as follows:

'Jubilee' class 4-6-0 No 5573 *Newfoundland*:
Blue/grey livery with full lining out in prewar style, but using maroon instead of black for the edging, with straw yellow lining and sans-serif insignia.
'Jubilee' class 4-6-0 No 5594 *Bhopal*:
Maroon livery, with one side lined in black and straw yellow, the other having plain maroon with only a single straw yellow line along the running plate. Splashers were plain maroon on both sides. Straw yellow sans serif insignia, without shading.
'Princess Coronation' class 4-6-2 No 6234 *Duchess of Abercorn*:
A *lighter* shade of blue/grey, with maroon edging and straw yellow lining and insignia. One side was more fully lined than the other, as in the case of *Bhopal*. The nameplate had a maroon background with the lettering picked out in straw yellow.

Two locomotives, Nos 5594 and 6234 were brought to Euston for official inspection, being housed overnight at Willesden roundhouse on 23 March 1946, where the well known photographer H. C. Casserley was fortunate to record them. These pictures appear as Plates 33b and 33c of D. Jenkinson's book *Locomotive Liveries of the LMS*, and he also reproduces an excellent colour sample of the blue/grey, complete with lining-out. Unfortunately Mr Jenkinson is mistaken in describing No 6234 as having a 'half and half' paint finish to the rear of the tender; what is seen in Mr Casserley's photograph is a trick of light showing the reflection of the roundhouse windows! I saw both locomotives some months afterwards, in general service, by which time a goodly layer of dirt had all but disguised their novel colour schemes.

One influential board member at this period of LMSR affairs was R. A. Riddles, then the vice-president for engineering, and he has since personally told the author of his stated preference for black as a locomotive livery; on more than one occasion. Perhaps it was Riddle's persuasion, or perhaps just an act of bowing to the inevitable that led to the final choice of black as the postwar locomotive livery, commencing in the early summer of 1946.

The broad specification was as follows:

Locomotives

Passenger livery

Overall glossy (varnished) black finish, with maroon* edging and single straw yellow lining. To begin with (1946), the insignia were in plain straw yellow bold — 'grotesque' — sans-serif; hand painted. Bufferbeams were vermilion red, without lining. The nameplate was *painted*, with a maroon background and straw yellow letters and border. (Note that the sans serif letters and numerals were *not* Gill sans, as has been frequently mis-stated, but 'grotesque', which is a sans serif variant.)

Within a few months, two changes were made to this scheme. One was that on the running plate valance, from front to rear, the maroon was lined on *both* sides with straw yellow (forming a bold band of colour) and two boiler bands were painted to match — the one nearest the smokebox and the one nearest the firebox, plus the band where the firebox met the cab front. The other change involved the insignia, which had a fine maroon line added within the straw yellow, (see illustrations). Some smokebox numberplates were still produced in the '1936-style' sans serif form for this livery, but the majority had serifs.

Mixed traffic and goods engines

With the exception of all the Stanier 'Princess Royal' and 'Princess Coronation' 4-6-2s, the 'Royal Scots', 'Patriots' and 'Jubilees' (in rebuilt and original forms), and the solitary, unique 'Turbomotive', *all* locomotives were to be in plain black, with a vermilion red bufferbeam without lining; which meant that there was no longer any differentiation between mixed traffic and goods types, in so far as their livery was concerned. The same style of lettering and numerals, in straw yellow with inner maroon line, was used as on the express passenger engines, with a smaller size of numerals for some types.

It should be emphasised that comparatively few locomotives in this lesser category received the new livery prior to nationalisation. The most numerous being the new mixed traffic Class 4 and 2 2-6-0s and the Class 2 2-6-2Ts constructed to H. G. Ivatt's design during 1947/early 1948.

**I have not given a separate colour sample for maroon in this book, because no official statement that it was meant to differ from the crimson lake has been discovered. Personally I think it was more purple, or bluer; a matter of opinion.*

Diesel locomotives

Plain glossy black with aluminium/silver insignia and trim; aluminium paint for bogies and roof for main line types.
Plain black with straw/maroon insignia for shunters.
Vermilion red bufferbeams on the latter.

Passenger Stock

Maroon.
Straw yellow lining, with straw/black/straw at waist (late prewar style). Grey roof.
Black ends, bogies, underframes, etc.
Electric stock was in similar livery, but in some cases only lined with a straw/black/straw line at waist level.

Other Stock

Most non-passenger carrying stock capable of working in passenger train consists was painted in plain maroon (as the late prewar style) but some passenger brake vans had lining to match the carriages. Roof and other details as above.

Goods Wagons

The wartime austerity scheme was for the most part continued, with the minimum of bauxite paint on wooden construction. No basic change from the late 1930 livery style for repainted or new vans.

66
The outbreak of World War 2 put paid to the LMSR streamlined trains and later also to the crimson lake livery. Overall black became the order of the day, and the streamlined casings that were to have graced the new Stanier Pacifics Nos 6249-6252 built in 1944 at Crewe were abandoned, but not before the tenders for them had received their streamlined fairings! Previously Nos 6245-6248, built at Crewe in 1943 had been delivered in streamlined form but painted plain black. No 6252 *City of Leicester* shows the wartime livery, devoid of all lining-out and with yellow and red numerals and insignia, and vermilion red bufferbeam. It has been stated that these engines had red backgrounds to their polished brass nameplates, but this photograph clearly shows a black background. The boilerside handrail has been left in polished steel; all others being painted black.
British Rail

66

67

68

67
The Stanier '8F' 2-8-0 was chosen by the War Department for quantity construction, for use overseas and on the home railways, and the workshops of the LNER, SR and GWR all constructed batches, as well as the LMSR and outside contractors. The plain black livery was in effect the same as the prewar goods engine livery, and was relieved only by the yellow and red numerals and insignia, as seen here on No 8400, the first example built at the Swindon Works of the GWR, in 1943.
British Rail

68
The wartime livery was perpetuated for some time after the restoration of peace, and No 2242, a Fairburn version of Stanier's Class 4P 2-6-4T, built at Derby in 1946 shows it well. Of interest is the smokebox numberplate, in the 1936 style! The Scottish enginemen have groomed their locomotive to perfection, with pale blue background to numerplate and shedplate and silver paint embellishments to the smokebox door and pony truck guardirons. A black engine *could* look beautiful! No 2242 was photographed at Glasgow Central in April 1948.
H. C. Casserley

69

69
The postwar LMSR livery experiments are described in some detail in the text, and I hope that some of the confusion relating to this photograph in particular has now been clarified. This picture purports to depict Stanier Pacific No 6235 *City of Birmingham* in de-streamlined condition and in photographic grey paint. It must be emphasised that it *never ran* like this, in particular without a new casting for the double chimney; what is seen here is the crude casting which existing streamliners had *beneath* the outer sheet metal casing. Coincidental to the decision to remove the streamlined casings from the Pacifics, the LMSR was seeking a new postwar livery style; the colour probably to be either crimson lake or a blue/grey shade, with simplified sans serif numerals and insignia, and lining. This picture shows one proposal with a fully lined and edged bufferbeam, but with no lined boiler bands and only a single yellow (pale straw) line along the running plate (see also next photo) and with yellow and black edging to cab and tender; a scheme suitable for either livery colour. In fact, it was No 6234 *Duchess of Abercorn* that was actually painted in the blue/grey, in March 1946.
Crown Copyright, National Railway Museum, York

70
Two 'Jubilee' class 4-6-0s featured in the 1946 livery experiments: No 5573 *Newfoundland* which ran in the blue/grey colour, with full prewar style lining-out in pale straw yellow and black but with unshaded sans serif numerals and insignia, and No 5594 *Bhopal*, seen here, which was in maroon. On this side of No 5594 only a single straw yellow line was applied, from end to end of the running plate. The other side was more fully lined for comparison, (but there was no lining on the splashers or boiler bands). The insignia and numerals were in bold sans serif straw yellow; note the close spacing for 'LMS' on the tender, and no lining on bufferbeam. *J. H. Platts*

70

71
The outcome of the livery experiments was a decision to retain black! A glossy finish was favoured, with straw yellow and *maroon* (this was the official description), lining and edging was adopted for express passenger locomotives. At first the lettering and numerals were in plain straw yellow, but an inner maroon line was soon added, (see drawing). 'Royal Scot' class 4-6-0 No 6134 *The Cheshire Regiment* displays the new livery in this official broadside. Note that all the handrails, the wheel rims and bosses were painted black; only the valve motion and nameplates had a polished metal finish. The new livery soon included the addition of lining to the boiler bands at each extremity and to the firebox next to the cab. *British Rail*

71

72
This view of the unique Stanier 'Turbomotive' 4-6-2 No 6202, seen leaving Crewe with the 8.30am Euston-Liverpool train on 14 June 1947, shows that the rear of the tender was in plain black. Another feature of the postwar black livery was that the power classification was placed *below* the number on the cabsides on the larger types of locomotive. *H. C. Casserley*

73
H. G. Ivatt continued Stanier's policy of rebuilding the Fowler express passenger 4-6-0s with taper-boilers, and both the 'Royal Scot' and some of the 'Patriot' class were so altered in the final LMSR days (and afterwards by BR). The lined black livery sat quite smartly on these excellent machines, as is shown in this view of 'Rebuilt Patriot' 4-6-0 No 5526 *Morecambe and Heysham*; photographed in 1947 at Camden shed. In the lefthand background a portion of the cabside of 'Princess Coronation' class 4-6-2 No 6223 *Princess Alice* can be seen, with the numerals in plain straw yellow, ie without the maroon inner line. The nameplate on No 5526 has a maroon background to the polished brass.
P. Ransome-Wallis

74
Photographed on the occasion of the naming ceremony in his honour — 17 December 1947 — Sir William A. Stanier FRS is seen at the controls of No 6256, one of the two final 'Princess Coronation' Pacifics introduced by H. G. Ivatt, with certain modifications. The cabside numerals are well illustrated in their sans serif 'grotesque' style (*not* 'Gill sans') showing the maroon inner line on the straw yellow. Also evident is the maroon edging to the cabside, separated from the black by a straw yellow line. *British Rail*

72

73

74

75

76

75
For mixed traffic locomotives, Ivatt retained a plain black livery overall, but with the maroon inner line added to the straw yellow sans serif lettering. His new Class 4F 2-6-0 No 3001 was photographed at Bletchley in April 1948, complete with the hideous original double chimney. Note that the power classification '4F' is carried *above* the numbers in this instance. *P. Ransome-Wallis*

76
No 6419 of the smaller Ivatt mixed traffic 2-6-0 design, the Class '2F' is seen here, at Manchester Victoria & Exchange station on 24 April 1947, in the plain black livery with new insignia. One curiosity was that the larger 4F design had '1936-style' sans serif smokebox numberplates, whereas these smaller engines had the normal serif version. *H. C. Casserley*

77
Perhaps Ivatt had memories of No 5552 *Silver Jubilee* in the special black and chromium livery, (see photo 22) when he chose the striking black and silver (aluminium) livery for his first main line diesel-electric locomotive No 10000, which appeared in December 1947, just three weeks before the LMSR became a part of the new nationalised 'British Railways'. The lettering and numerals and the line around the body at waist level were in relief polished aluminium, and the roof and bogies were painted to match. White headcode discs were carried instead of oil lamps during daytime. No 10000 is seen here soon after entry into service, heading a Midland line express. *E. R. Wethersett*

78
Diesel shunting locomotives were always in plain black livery, to begin with yellow and red lettering and numerals were applied but under the postwar livery scheme they were intended to receive the straw and maroon sans serif characters. No 7080 is seen however in the livery it received in 1939; when delivered to traffic. *Ian Allan Library*

77

78

79

$12\frac{1}{2}''$

LMS

$10''$

12345
67890

79
The 1946 insignia, destined to last only 18 months, until the LMSR was nationalised; a bold sans serif style in pale yellow (straw) with an inset line of maroon, except for some early examples which were in plain straw. The height varied from $12\frac{1}{2}$in (as shown) to 14ins, for the letters, and from 10in to 12in for the numerals.

80

80
The crimson lake livery of prewar days, with the simplified lining style was restored for postwar construction, with straw yellow replacing the chrome yellow; also on routine repaints. The roof was grey and the carriage ends were black, together with the underframe, bogies and buffers. Two small detail changes should be noted (compared with photo 49): the third class designation has a redrawn 3 with flat top, and the running numbers were in serif style, to match the 'LMS' insignia. It is interesting to note that the circular crest was used throughout from 1923-1947 on some types of main line passenger stock. Illustrated is third open No 27106. *British Rail*

81
First class kitchen/dining car No 43, of prewar build, was refurbished in 1946 with a modern interior. At the same time it received a modified livery, retaining the crimson lake (or maroon) but with straw yellow and black lining and straw yellow sans serif letters and numerals. These were clearly intended to match the new black locomotive livery, but only a few examples were so painted before nationalisation. The crest had straw yellow instead of chrome yellow for the letters and surround. Note the change of title to 'Restaurant Car', instead of 'Dining Car'. *British Rail*

81

82

83

82
During the war period all lining-out was discontinued if a carriage received a full repaint (most it seems were only patch-painted, for economy) and this Southport-Liverpool train of compartment stock, seen near Seaforth, is in the plain maroon livery. The later style of flat-topped 3 is used on the doors; the buffer beam on the motor coach was in vermilion red. *W. Hubert Foster*

83
The postwar livery for electric rolling stock had the simplified lining, in this instance the carriage sides have a single broad band of black along the waistline beading, edged by straw yellow lines. No crest is carried and buffer beam area is black on this type of stock. Photographed at Manchester Victoria. *W. Hubert Foster*

84

84
The war brought about an austere livery change for wooden-bodied goods wagons. New construction had virtually no paintwork on the wooden portions, with just patches of bauxite where white lettering was needed. Metal details remained in bauxite, including the solebars, buffer shanks and headstocks. Wheels and other details below the solebar were black. Note the very small lettering for LMS on the top example a 13ton open with steel ends, whilst the lower one is an engineer's departmental wagon, with 'Stoke' district, but not marked LMS.
Crown Copyright, National Railway Museum, York

3: Stations and Buildings

Unlike the GWR and the SR, where the civil engineers had early on prepared a fairly rigid specification for the colours to be used in painting stations and buildings, the LMSR seem to have retained the varied schemes of their pre-Grouping constituents for the first decade or so, although the new namestyle 'London Midland and Scottish Railway' appeared on signs, and poster boards; usually in black and white. Perhaps this was because the various colour schemes were actually quite similar, or perhaps it was a matter of economy; using-up existing paint stocks. However, in the mid-1930s a list of colours was decided upon, but its exact application to each structure still seems to have been vague, and left to the discretion of the local engineer, or the foreman painter. The basic rule seems to have been to 'pair' the colours, one light with one dark. The dark colours generally being on the lower portions of structures, cast iron columns, doors, ironwork of seats, etc. (Black was sometimes substituted on small items of ironwork.) Above the dark colour, the lighter one was used for the upper part of wooden wall surfaces, roof and awnings and upper part of some columns, also window frames and sashes if these were not white. The listed colours for stations were as follows:

Dark: Middle brown/Venetian red/Mid-Brunswick green
Light: Deep cream/Portland stone.

Use of the green appears to have been limited to the more rural stations, and what scant contemporary colour photographic evidence the writer has unearthed shows the venetian red and deep cream as favoured for larger stations and the buildings in important goods yards, etc. For stations on electrified lines a Golden brown colour was sometimes applied in order to mask the effects of the brakedust from these trains. This was either used alone, or with deep cream. Poster boards, signs and nameboards were basically white on black, using raised letters (many dating to pre-Grouping days) except for the warning signs which had black letters on white, as a rule, although red on white was to be found on some trespass signs and some public crossings; certainly in postwar days.

In the 1930s the LMSR produced a new design of station nameboard known as the 'Hawkseye' target type (see drawing) and at the same time the small signs hung adjacent to lamps were in black letters on yellow enamel. In the postwar period some stations received a new style of sign, with white Gill sans on a maroon enamel background (pre-dating the very similar BR type!) and one of these is seen in the photograph of Watford

85

85
Until the mid 1930s the LMSR seems to have been content to retain many of the pre-Grouping colours and station signs and furniture that it had inherited. Thus this scene at Crewe differs but little from LNWR days, although the hanging signs have been freshly painted in white on black. The letters, numerals and the 'pointing finger' direction sign are all in raised characters, screwed to the wooden backing frame. 'Royal Scot' class 4-6-0 No 6146 *Jenny Lind* is seen arriving, on an evidently well patronised express duty.
Real Photos

Junction (photo 92). This was probably a future standard type.

A more rigid painting specification existed from 1931 onwards for signalboxes, using light stone and dark brown, as follows:

Dark brown: Doors, staircases, guttering pipes and facia boards, barge boards, bottom sills, corner posts and window cleaning stage.
Light stone: All other areas of woodwork except window frames and sashes which were white.
The signalbox name was specified to be in white letters on black; firebuckets in red.

It has been recorded that signalboxes on the Central Wales line were painted green and cream in 1937; possibly this was because of their rural location (just as some stations were painted green and cream in country areas).

86
The attractive station at Bedford St Johns, formerly LNWR, seen in LMSR days. Poster boards and 'Gentlemen' sign in black and white, upper part of canopy, with decorative cast iron support, and fencing, probably in deep cream or portland stone; cast iron columns and doors probably in venetian red or middle brown; although sometimes these rural stations were painted in mid-Brunswick green. *British Rail*

87
A similar colour scheme in either the venetian red, the brown or the green, would have been applied to St Albans Abbey station with the slender cast iron columns in venetian red, brown or green and the upper ironwork and woodwork in deep cream, or portland stone. The Stanier Class 2P 0-4-4T No 1908 stands with the 2.45pm push and pull train for all stations to Watford Junction, and is in plain black livery with yellow and red insignia. '1936-style' smokebox numberplate fitted. Photographed on 14 August 1948. *E. D. Bruton*

86

87

88

88
This view of the frontage of Kilburn station is of particular interest because it shows, above the entrance, one of the large hand-painted posters that the LMSR (and the LNWR beforehand) specialised in. Note the monogram made from the letters LMS, a feature of some publicity material but never adopted for rolling stock or locomotives. The station name sign, repeated twice, is almost certainly in black and golden yellow, whilst over the top of 'Williams Drug Stores' a shabby notice in relief aluminium Gill sans lettering reads 'Station Entrance', probably on a dark brown or maroon background. It seems that 'Kemps the Grocers' have succeeded in providing a more legible and simple sign than that advertising the existence of the railway station!
Ian Allan Library

89
Raised sans serif letters, painted white upon a black board were a common feature on LMSR station frontages — many dating from pre-Grouping days, but with the full title of the new railway added. This example, photographed in November 1949 (nearly two years after the LMSR ceased to exist) was over the road entrance to Hampstead Heath station. The name of the station has been erased, almost certainly during the wartime invasion scare, when it was thought that German paratroopers would be able to establish the whereabouts of their landing by reading station names and road signs! *British Rail*

89

90

Golden yellow beaded glass background
Black lettering
KENTON
Black surrounds
White panel
Creosoted frame

91
A variant of the 'Hawkseye' target nameboard was produced for the new stations on the joint LMSR/LPTB electrification to Upminster (District Line). These were either glossy paint finish, or perhaps enamelled, and the background colours were reversed, with the station name in black upon white and the outer surround with the wooden frame in golden yellow. This example was constructed to fit the top of the platform seat below.
Crown Copyright, National Railway Museum, York

91

90
The standard 'Hawkseye' station nameboard design, with a 'target' shape rather like that used by London Transport and by the SR; introduced in the 1930s. These nameboards were widespread on the LMSR system. (See also photo 109.) The golden yellow background to the lettering had a beaded glass effect to reflect light and to make the name itself more visually prominent.

92

93

92
Watford Junction main line station on 9 September 1946, with the branch to St Albans in the background, and two ex LNWR 0-6-2Ts Nos 6909 and 6725 in evidence. The new station sign in the lower lefthand foreground has white letters and surround on an enamelled maroon panel and white painted supports and frame; a new postwar style, using Gill sans lettering. (Note the abbreviation 'Jn'.)
H. C. Casserley

93
This standard small type signalbox at Aber, on the Chester-Holyhead main line, photographed in 1937, shows the standardised paint scheme of light stone and dark brown.
LGRP courtesy David & Charles

4: Road Vehicles

The LMSR had a very large fleet of road vehicles, ranging from horse-drawn carts to buses and coaches, with of course delivery vans being the most commonplace. The company placed considerable reliance upon the cart horse until the end of its days, and it was BR that quickly got rid of the large stud it inherited; replacing them with 'mechanical horses'. From 1928 onwards omnibuses were an important feature of operations; following the passing of the 1928 LMSR Road Transport Act, which enabled the railway to operate competitively on the roads, as well as the rails. A selection of road vehicles are illustrated, to give a basic idea of the liveries carried at various times.

Buses

Basically in crimson lake livery, with gold (gilt) lining and lettering; black mudguards and trim; white roof. However variations were commonplace, with cream used for bodywork above the waist, and with red lining added to the gilt. LMSR crest sometimes carried.

Horse-drawn vehicles

These featured crimson lake for the main wooden bodywork and wheels, but black was used for fabric hoods and weatherproof protection A mixture of lettering, some in white sans serif or serif, and some in yellow or gilt serif form was used. (The white version usually being on the black portions.) Lining was in gilt on the crimson lake.

Motor vehicles

These followed much the same basic style as the horse drawn vehicles (which they were developed from) with crimson lake as the main body colour and with black hoods, mudguards; etc. As a rule lettering was sans serif or serif white on the black areas and serif gilt, or yellow on the crimson lake. A 1930s development was the use of red shading to the yellow insignia (which became a deeper chrome yellow, resembling 'old gold').

NB In the postwar period all types of road vehicles assumed a plain maroon livery, with white lettering and black trim; it is probable that the maroon paint originated in wartime, just as in the case of the passenger rolling stock; it appeared to be more opaque and slightly bluer.

94

This Leyland Motors 'L' or 'Lion' type bus was introduced in 1925, with pneumatic tyres and a purpose-built chassis (not the usual adaptation of a goods vehicle chassis). The LMSR received this example, resplendent in crimson lake, with gold and red lining-out and black beading. The wheels were in crimson lake with a fine gold line and polished metal hubs. The roof and roofboard were in white, with the letters 'LMS' in black. Crest applied to bodyside. *Ian Allan Library*

95

This superbly elegant Albion six-cylinder 24-seater bonnet-type coach for the LMSR was exhibited at London Olympia in 1929. The livery is crimson lake for the main body and roof, with gold lining and gold sans serif lettering, shaded black. The bonnet, the mudguards, the driver's windscreen surround, and wheels were black. The gold line continued along the bonnet to the chromium plated radiator. *Ian Allan Library*

94

95

96

96
The 'Karrier Ro-Railer' was a road vehicle adapted to run on both road and rail, and was tested on the LMSR Harpenden-Hemel Hempstead branch in the winter of 1930/31. Afterwards it went for a while to Stratford-on-Avon to work a service to the LMSR Welcombe Hotel. The livery was crimson lake with a white roof, gold lining to the beading and gold serif lettering shaded black. The front mudguards and the wheel centres were black. This curious vehicle ended its days in engineer's service on the LNER West Highland line. *British Rail*

97
The LMSR had the largest stud of horses in the country, and until its final days it was an intensive user of horses. This line-up of horse drawn vehicles for city deliveries shows the white sans serif lettering used on the front weatherboard (where fitted) whereas the letters 'LMS' on the sides (wherever a covered body was used) were in serif form in white on the black canvas hood. The lower part of the bodywork (wood) was finished in crimson lake. The vehicle on the extreme right of this picture carries the legend 'Express Parcels Traffic'. Wooden drays (flat wagons) had the full title 'London Midland and Scottish Railway Company' in white upon the crimson lake side raves, and the letters 'LMS' on the end raves; together with the number. *Real Photos*

97

98
A photograph taken in 1923, and therefore showing the earliest LMSR livery for horse-drawn vans. The main body and the shafts and wheels are painted in crimson lake, with pale yellow (to resemble gilt) lining and lettering. Note that the shading effect of 'LMS 23' is *all* in yellow, as is the italic script giving the name and address of the railway company. Inner black band to the lining on the wheel rims. The hood was black with white lettering. *Crown Copyright, National Railway Museum, York*

99
A 1929-built Karrier Motors Ltd lorry, with white sans serif lettering and numerals. Basic livery of crimson lake and black, probably with yellow lining. Note the solid tyres fitted and the canvas roof to the cab, which has no windscreen and a rolled tarpaulin to protect the driver in bad weather. *Ian Allan Library*

98

99

100

100
A Karrier 'Bantam' van for express parcels traffic, delivered to the LMSR in the summer of 1934. The bodystyle still owes much to its horse-drawn predecessors but the driver now has a fully enclosed cab, and pneumatic tyres are fitted. The livery appears to be the same as for the horse drawn vehicles, with the fabric upperpart in black with white lettering and the main body in crimson lake with pale yellow (gilt shade) lining and lettering. *Ian Allan Library*

01

101
A picture that speaks for itself! The container was in crimson lake, with deep chrome yellow (or possibly gold leaf) lettering; complete with LMSR crest. The side rave on the lorry has white lettering and numerals, and the rear tyre wall has been painted white. All the lettering on the container is shaded to the right and below in black. *Ian Allan Library*

102

103

102
A rear view of one of two Leyland 'Cub' horse boxes, with excellent bodywork by Vincent's of Reading, supplied to the LMSR in 1934. White roof to driving cab and horsebox; plain crimson lake livery with black mudguards and fuel tank. The lettering is in deep chrome yellow with red shading to the left and below for LMS. Crest on rear door. *Ian Allan Library*

103
The 'Mechanical Horse' was the designed replacement for the faithful four-legged animal — and only required *three* wheels! This 6ton example was photographed in October 1941, complete with wartime white paint additions to make it more conspicuous in the blackout. Basic livery of crimson lake (perhaps maroon by now — see text) with white sans serif lettering. The female driver (another wartime feature) wears an LMS white metal badge on her cap. *Ian Allan Library*

5: Miscellany

A few associated elements in the overall livery of the LMSR are shown, in order to give added detail. However, the reader in search of 'in-depth' descriptions for which space does not exist in this series, must search through contemporary publications where quite often a description of a new ship, for example, will have some limited livery detail included. For the printed publicity of the prewar and postwar periods, a useful source can be found in the specialist journals of the day produced for the commercial art world. Much information on uniforms can be gleaned from the study of people in the backgrounds to photographs taken at stations, despite the fact that these nearly always concentrate upon the locomotive and train as their main subject. Recently a splendid book entitled *LMS Miscellany* by H. N. Twells (Oxford Publishing Co) has appeared, giving much fascinating background material to the day-to-day operations of this great railway; it is thoroughly recommended as a source of much rare and accurate historical detail.

104
For the period 1925-1947 the standard livery for LMSR ships was a black hull with white/varnished teak upperworks, and a buff coloured funnel with a black top. Between 1923-1925 there was a red band between the buff and the black, and this was retained for the Goole based ships, which also carried the letters AHL (Associated Humber Lines) on the red band from 1935 onwards. The LMSR House Flag for ships had a horizontal white cross on a red (perhaps crimson lake) ground and an LMSR device in the centre of the cross in a circle. Illustrated is the ill-fated steamer *Princess Victoria*, built for the LMSR by Denny Bros Ltd in 1947. Six years later she was sunk by a heavy gale, off the County Down coast, with the loss of 128 lives, when the extremely rough seas burst open the inward-opening rear doors used for loading and unloading road vehicles.
British Rail

104

105
The publicity department sometimes made a monogram of the letters LMS (see also photo 88) but it was never properly adopted, unlike that created for the GWR for example. This mileage board at Wick station shows one early version; photographed in July 1931.
H. C. Casserley

105

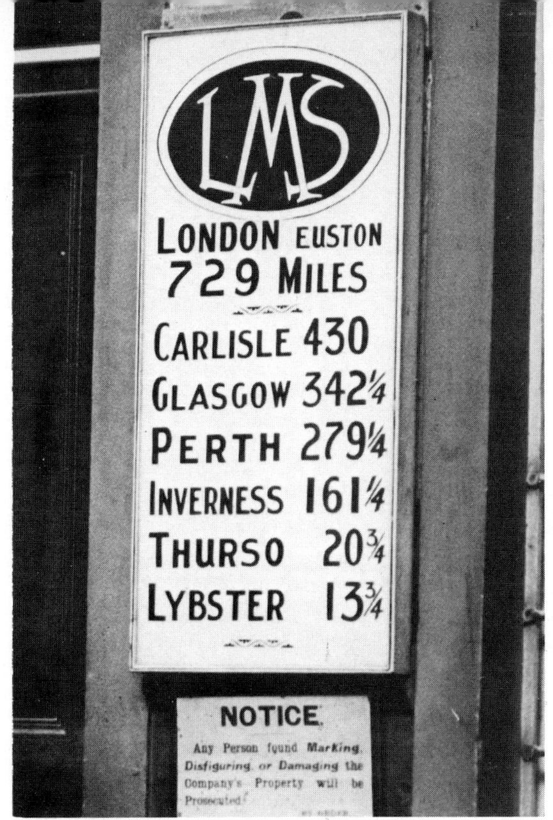

106

107

106
Before the arrival of the neon sign upon the streets, large advertising displays were created by lamp bulbs, arranged to flash on and off in sequences, and in differing colours. This LMSR display in Leicester Square, London in the late 1920s, had the lights arranged to make the wheels and valve gear on the locomotive appear to move, also the smoke! (The price of a luncheon or dinner cannot go unnoticed!) *Author's collection*

107
Two typical LMSR posters of the late 1920s, in a style which carried on until 1947, with the use of many famous poster artists, as well as some Royal Acadamecians. The poster for the hotels is by Horace Taylor, and the fishing scene is by Norman Wilkinson RI, who did a magnificient series on this Sport theme, as well as many for the LMSR shipping services.
Author's collection

108

108
Two top-link Camden enginemen look at the Engine Arrangements board at the shed, which shows them which locomotive is rostered to them for the day and which train they will haul. The uniform of the driver was typical of the LMSR period and it is interesting to note that both he and his fireman (left) have white shirts beneath the overalls!
Author's collection

109
Although the tender of the 'Princess Coronation' Pacific is lettered 'British Railways' the locomotive is still in the final LMSR black livery with maroon and straw lining. No M6230 *Duchess of Buccleuch* pounds through Berkhamsted station on 25 March 1948; three months after the LMSR ceased to exist. The picture is included because it shows several distinctive LMSR features: the 'Hawkseye' station nameboard, angled at the platform end; the black and white named train board carried on the leading carriage of the 'The Royal Scot', and the LMSR crest and final livery style on the carriage itself.
H. C. Casserley

109